Yellow Umbrella Books are published by Red Brick Learning
7825 Telegraph Road, Bloomington, Minnesota 55438
http://www.redbricklearning.com

Library of Congress Cataloging-in-Publication Data
Trumbauer, Lisa,
    Water/by Lisa Trumbauer.
    p. cm.
    Summary: "Simple text and pictures present the different forms water
can take"—Provided by publisher.
    Includes index.
    ISBN-13: 978-0-7368-5977-6 (hardcover)
    ISBN-10: 0-7368-5977-2 (hardcover)
    ISBN 0-7368-1709-3 (softcover)
    1. Water—Juvenile literature. I. Title.
GB662.3.T78 2006
551.57—dc22   2005025741

Written by Lisa Trumbauer
Developed by Raindrop Publishing

Editorial Director: Mary Lindeen
Editor: Jennifer VanVoorst
Photo Researcher: Wanda Winch
Developer: Raindrop Publishing
Conversion Assistants: Jenny Marks, Laura Manthe

Photo Credits
Cover: ChromaZone Images/Index Stock; Title Page: DigitalStock; Page 4:
ChromaZone Images/Index Stock; Page 6: DigitalStock; Page 8: DigitalStock;
Page 10: Kent Knudson/PhotoLink/PhotoDisc; Page 12: Paul Hartley/Image Ideas,
Inc.; Page 14: Larry Larimer/Brand X Pictures; Page 16: PhotoLink/PhotoDisc

1 2 3 4 5 6 11 10 09 08 07 06

Water

by Lisa Trumbauer

Yellow
Umbrella
Books
for early readers

Rain is water.

Snow is water.

Ice is water.

Hail is water.

Dew is water.

Steam is water.

Fog is water.

Index